Project Manager Productivity Hacks

2021 Edition

by

Nigel Creaser

two7four

PUBLISHING

If the eagle-eyed among you spot a typo, please let me know by dropping me an email with the subject *Typo Alert* to typo@two7four.com.

Contents

Introduction 5

Notifications 9

Schedule Email.................. 25

Outlook Rules!....................41

Do Not Disturb................... 59

Mobile Subservience 65

Outlook Quick Actions 73

OneNote 79

Delegate 87

Prioritise 101

Remote Working 115

Be a Dictator...................137

Thank You 145

Dedication..................... 149

Also By 151

Nigel Creaser...................151

Coming Soon 159

About the Author............. 165

Introduction

Hi, and welcome to Project Manager Productivity Hacks.

Why write another book on productivity hacks? Are there not enough of them already? The trigger my interview with Sarah M. Hoban for my podcast The Sunday Lunch Project Manager Podcast. (plug, plug)

https://anchor.fm/sundaylunchpm/episodes/meets-Sarah-M-Hoban--PMP--The-Productive-PM-Part-1-e15gjv

We talked about improving productivity for project managers and during the conversation I recalled that a while ago I started a blog post in which I planned to share some ideas that I have used in my career, to improve my productivity and get out of the office that little bit earlier. Over the years I have shared these with others and almost every time I received a positive and enthusiastic response.

Having rekindled the idea, I sat down and noted 11 hacks, tips, shortcuts, whatever you want to call them, that I have used over the years. Why 11? Well, it's one more than 10. This book goes up to 11. (One for the Spinal Tap fans.)

I see productivity as remarkably similar to an exercise programme, we all start off with good intentions and some of the changes work great for us and others don't. Some changes stick and become part of our daily life whereas others are just a real chore. Even the ones we found that work well for us and are straightforward to implement, can, under certain circumstances, end up being dropped by us. Sometimes this happens consciously and sometimes little by little over time.

Whether building healthier habits or improving our productivity it's not just a case of making the changes and we are done. It takes repeated work. If you find them hard to implement, then they may not be for you in your current situation. If you start using them and end up letting them drift, then start again.

Improving our productivity is something many of us strive to achieve aiming for some kind of efficiency nirvana. Let me tell you, it ain't gonna happen, perfection is a myth. Give yourself a break. It's a bit like yoga, you can't win at yoga, no matter how bendy you get you can always be bendier, that's the same with personal productivity, you can't win at productivity, you can always be productivier (Ed. Is that a word?)

I have not always been able to apply these hacks successfully, but that is not a reflection of the hacks themselves, it is more about my ability to stick with them even though I know they work.

One last point, as with your fitness regime trying to make loads of changes at once can spell disaster. I recommend that once you have a read of these hacks, choose one that excites you (maybe not excites but seems to resonate) and have a go at it. Once it is second nature to you, move on to the next one that spoke to you. If you find that the first one was not for you then bin it. Choose another and try that one out, then rinse and repeat.

So, on with the hacks...

Notifications

I have one simple piece of advice when it comes to notifications:

"Turn the bloody things off!"

Is that clear enough? We don't need the Dom Joly 1990's, cool to have, Nokia ring tone (Dom Joly is a UK Comedian who was famous for wandering around in one of his sketches with a massive phone that would start ringing loudly in the most inappropriate places, like a cinema which he would answer shouting about where he was and disturbing all those around him.) and all the beeps, pings and vibrations, that have arrived since.

Why are you so bothered about this? I hear you ask. There was a study completed by Professor Gloria Mark from the Department of Informatics at the University of California, Irvine where they found studying some 400 people that when they got interrupted, each interruption cost something like 40 seconds. That does not sound like too much, does it?

The most important part of this was that the 40 seconds was only the actual physical length of the interruption not the actual cost to the productivity in that day. They estimated that it took around 23 minutes for someone to get back to the same level of focus that they were on before that interruption.

I am not sure whether 23 minutes is an accurate figure, it seems long to me, but even if it was ten minutes, five minutes, three or two minutes when you multiply it by the number of interruptions you get from people each day, it is significant.

Think about the number of interruptions we get from each of the tools we have that are trying to, in Joey air quotes, help us (In many cases they are there just to attract your attention to demonstrate their worth rather than in your best interests). Even if they do not fully derail you, they still take away your focus.

How many times do you think *you* get interrupted in a day? Ten? Twenty? Thirty? Two hundred? If we went low and said there were ten minutes lost, so it's easy maths. If it cost you 23 minutes that's a total loss of 230 minutes, that's nearly four hours, that's massive.

Are some of you are thinking, like me, that not all of those 23 minutes will be entirely unproductive time? It is a fair point. Even if we said that of the 23 minutes that 3 minutes ended up "lost" and we take our 10 interruptions from before, when multiplied out that costs you 30 minutes a day. Is that your lunch, or your commute back? Is that worth saving? It is to me, if it is to you then keep reading, otherwise, you may have grabbed the wrong book.

Shut Up Email

If you are still with me then let's take a look at how we can deal with your email as a first step. One of the most intrusive tools we use is email. It dominates our working life and many of us have hundreds of them each day. Having a notification for each one can cause a level of interruptions that impacts heavily on our daily productivity. I am going to use Microsoft Outlook as the

example here because that's what I've
seen most people use.

When we receive an email, Microsoft
tells us. That's very kind of them. It is a
bit like the days when we would have been
sat at home and the letters drop through
our letterbox or when we were sat in an
office processing paperwork. A letter or
memo would be dropped into our in-tray.
The result was that we immediately know
we had a communication. We would see or
hear it and we would pick it up and deal
with it.

Nothing has changed then; the only
difference is it's a computer or phone
telling you rather than your letterbox
or in-tray now. If you were away from
home or away from your desk, then you
would be blissfully unaware of it.

Not quite! Back then what did we have?
One? Two? Five? Ten? Even 50 a day to deal
with, now we have thousands, literally
thousands of emails to deal with. (OK, not
literally, but a lot.)

Before the use of electronic email,
creating any correspondence took time

and effort, so it created a much higher barrier, now it's a few keystrokes and the whole world gets an email from you and more importantly to us now, vice versa. The fact that you received something previously meant it was more likely to be important and in need of attention, now it could be anything.

Outlook is not content with letting you know that you have a new email by incrementing the unread number in your inbox it feels that the email is so important that it should make a noise so you get the ping sound notification just in case you missed that. To be safe though, in case you were talking and not listening you get the little bit of toast appearing in the right-hand corner of your monitor. It is called toast because it pops up just like a toaster pops up your toast. That's not a diversion though, you're not going to do anything about it, are you? So finally, your mouse pointer flickers just in case you missed all of these alerts.

Does that seem a bit excessive to you? It does to me. When you install these programs, they have all of this up and running as default, so if you do not root around a bit it can be quite difficult to switch them off, but it can be done.

When I am sat next to some of my team members who have all of this switched on it seems odd. I have had it all switched off for so long it seems weird that it's on there. The toast is the one that I find the most intrusive because whilst you know you have an email from the ping, you know you have one for the icon and you know you have one for your mouse pointer flickering you don't know what the topic is or who it is from.

When it pops up, it makes you think "I wonder if that's the email from so and so?" or "Is it an email from the boss wanting something urgent?" or "The one from HR confirming my promotion".

The likelihood of you being diverted from what you are doing is lower than other notifications, it still distracts though. If we consider my earlier point about the time it takes us to get back into the flow state even an eighth as intrusive will cost us 30 minutes.

The toast makes it even more intrusive because it gives you that little bit of information that's like a teaser trailer at the end of an episode of your favourite show or clickbait on those annoying websites it tempts you to click.

It may show it's from the boss, it may have the title of your latest project or decision you have been waiting for, it might have important in the title.

All these things will make you want to go and look at it even more than the ping. Even if you ignore it, you now have it sitting in the back of your mind going "I wonder what the boss wants from me, should I check?"

The thing is nine times out of ten you go to look at it and it will be "Make sure that you clean the kitchen when you have your cup of coffee and return the teaspoons" or something with a similar level of criticality.

Having checked it, you have interrupted that business-critical task that you had just hit your flow state on and now that flow state has gone, it's broken and if the study referenced above is right then it will be 23 minutes for you to get back to that state.

Hopefully, I've convinced you that the email alerts will cause you problems. I turned them all off and I urge you to switch off yours too.

I am not looking to create a detailed instructional text here, especially seeing as the last time I authored an instructional book both of the damn software vendors changed the way that everything worked just as I finished my final review, so I had to redo it 😔. Having said that, the current steps at the time of writing, are as follows:

Open **Outlook,** click **File,** then click **Options** choose the section that says **Mail** and if you scroll down a little you will find something that says **Mail Arrival.** You now have four options, three of which are ticked, untick them **ALL** is my recommendation. You can then choose when to check and respond to emails.

Let's Be Social, Or Maybe Not

Email is not the only culprit, there are many other notifications from communications tools that distract us:

- phone text messages.

- social media likes, comments, shares and recently stuff that the social media giants have been paid to notify you about.

- Instant messaging, how many of you have had 15 open on your desk, typing in one thinking it is being sent to someone else, with embarrassing results.

- Slack channels.

- Teams channels.

- WhatsApp groups.

In recent years the number of additional channels which have been created seem to have exponentially increased at the office and in our personal lives. I have in recent weeks been in a situation whilst on a Skype call with my boss, I had

Teams message pop up, I had email open, and something popped up with urgent, my mobile phone started ringing, a WhatsApp message pinged on my phone and then vibrated on my watch, the house phone rang and then the front door was knocked on. I thought my head was going to explode!

I'm willing to admit that from a mental health point of view I have struggled with it. On that particular day, I was really feeling the strain and felt a little like a rabbit in headlights. The thing is, some of it was in my control, I had left my mobile phone volume on which I normally don't, I could switch off certain chatty WhatsApp groups, I could have set do not disturb on Skype, I could have closed Outlook.

This scenario does not happen all the time but every one of them is grabbing a bit of your time, they are diverting you as micro interruptions, they are so small you start to process them without thinking, then the important interruptions get missed. These tiny interruptions are like raindrops, one by one they are inconsequential, but get a lot of them together and it becomes a tsunami that devastates your hopes of being as effective and productive as you can be.

Who is in Charge of Your Time?

Our tendency is to leave all notifications on is driven by FOMO, the Fear Of Missing Out.

When considering whether you want to get past this, I urge you to consider the question above again.

"Who is in charge of YOUR time?

If not YOU, then who?"

When you decide to keep the notifications on you are handing over **YOUR** attention, **YOUR** productivity, **YOUR** effectiveness and **YOUR** health to who knows who.

ANYONE can send you an email. **ANYONE** can send you a text if they have your phone number. **ANYONE** can send you a Slack message if you are in that group. **ANYONE** can send you a Teams message if they are in your team.

The thing is, they don't know what you're doing at that point in time. They don't know whether what you are doing is more important than what they are sending a message about. Nine times out of 10 they don't even consider whether what you are doing is more important than what they are asking you to divert to.

Later I will talk about priorities, but understanding your priorities are of little use if you decide to hand it over your time to someone else.

I imagine some of you will be concerned about your VIPs. I understand that and have set up exceptions for those, where I can. For example, in Outlook you can have a pop up based on who the email is from even when you have all the other notifications switched off.

To be frank, in most people's jobs you do not need to respond or deal with the communication within minutes or seconds. If you are delivering on your priorities at work and by inference focussing on activities that support those priorities, then the fact that someone cannot get hold of you because you are focused on a piece of work that needs to be done then long term it won't matter, in fact, they won't even notice.

There appears to be an expectation that you have read every email or message as soon as it comes in. With the volume of emails received you cannot realistically do that; you shouldn't do that. When someone rings you up and asks if you read the email they sent you two seconds ago, the correct answer is "F*** **F!" Sorry I mean the appropriate answer is "Of course not, it only just arrived."

If you're in a role where there may be life and death scenarios, then be careful what you change. You will still need to divide the wheat from the chaff though. Review all your communications channels and switch off all but the critical ones, and for them, it may be a good idea to increase the level of the attention-grabbing features so they stand out.

Calendar Notifications

I have not mentioned calendar notifications. I will talk about using your calendar for scheduling your work in a later chapter. For now leave notifications for meetings and appointments on, though I would pick one method for the alerts, rather than having multiple things popping up.

Schedule Email

We all have days when we seem to be on email all day, and it can feel like every day. To get away from this, I recommend dealing with email only three times.

Only three times a day? Are some of you experiencing a slight twitch under one of your eyes, or at the corner of your mouth? Has your heart rate just increased, your smartwatch giving you an alert? Have you broken out in a cold sweat? Got sweaty palms? I know it feels decidedly uncomfortable, I felt the same way, and still do.

You may feel you need to do it more frequently and I have felt the same. What I have found is that by doing email more often I fall into a trap. The trap of feeling like I am getting stuff done, being the elusive superhero multi-tasker.

"What I really become is a common or garden unfocussed procrastinator."

By dealing with that email that has just popped in after checking my inbox ten times in the last twenty minutes (obviously notifications are now turned off) I feel "busy". The problem is that though it feels good to be busy, all I have done is interrupted what I was working on, or delayed starting that gnarly or boring task that I really do not want to do.

This goes back to our discussions about focus. The principle I am proposing is:

"When you are doing email.

Only do email."

If you have set aside time in your calendar to do your email and you keep checking your Slack channel or LinkedIn updates you will not be getting your email done. Keeping your time focussed is key.

Try this approach:

- Block out 30 minutes, 3 times a day in your diary.

- Use the 4Ds method for dealing with email. Delete, Delegate, Diarise or Do it.

30 Minutes

Why do I suggest 30 minutes?

Number one is it is what I heard as an approach many years back, I have tried it and it seems to work.

Number two is that it feels manageable but still creates a sense of urgency. If we do not constrain the time allotted to deal with email, we lose that crispness in our decision making. There is also the principle of Parkinson's Law which asserts that work expands to fill the time available. Parkinson's Law is named after Cyril Northcote Parkinson, who wrote what was intended to be a humorous essay for the Economist magazine in 1955.

https://www.economist.com/news/1955/11/19/parkinsons-law.

"If we have an hour to deal with emails,

we will take an hour.

If we have 45 minutes to deal with email,

we take 45 minutes!"

3 Times a Day

Why do I suggest 3 times a day?

Number one is the same as above, I have done it and it seemed to work.

Number two is that even though I have these slots in my diary I recognise that sometimes organisation priorities will require me to move them. Maybe requiring them to be moved later in the day or cancelled altogether. By having the other two slots there is still that time for dealing with email.

I have my slots in the early part of the morning to clear down any urgent items I need to do there and then or rearrange my diary, deal with a meeting invite and so on. My next slot is around lunchtime so anything key that has appeared since the morning session is only 2 to 3 hours old, a fairly reasonable time delay for an email response. My final slot is to wrap up the end of the day, spot things that need to go in my diary the following day and answer requests from people since lunch, again in a reasonable timeframe but not immediate.

The 4Ds Method

My approach when in the 30-minute slots is a modified version of what I have seen before, I have substituted the Defer category with Diarise as I feel the word defer is too open-ended and leaves us with a bunch of emails in our inbox with little flags next to them, using up our mental processing cycles.

What do I mean by the 4Ds method?

Every email needs a decision, which is why email is tiring. You have many, many micro-decisions to make when dealing with them. The 4Ds approach gives you structure and a way of prioritising them

Delete

My favourite. ☺ When you get into your inbox and find a whole bunch of stuff and then realising you do not care about it, there is nothing more satisfying than deleting it, in fact someday I select everything in the inbox and consider deleting them all.

Joking aside, this daft approach could be useful. Delete as the first item is not accidental. We need to make the judgement whether to address this email based on our priorities. If you decide that you are not going to delete it, then you are either going to use the company resources by giving it to one of your team or do it yourself, either now or later when you have diarised it.

"I recommend being ruthless,

if you cannot see any link to your goals

and objectives then you should not be

wasting your time dealing with it."

Delegate

I heard a statement that our role as a manager is to maximise the utilisation of company resources as well as delivering on our commitments. This makes sense to me and I do not think anyone would disagree.

What we sometimes miss is that we are one of those resources that we should maximise. When we are deciding whether we should deal with a particular email we must consider whether someone else in the team, who either costs less or influences less organisational resources, should be asked to deal with it. Giving them a chance to develop their skills into the bargain.

Diarise

If an email cannot be dealt with immediately meaning in 2 minutes or less or does not need to be dealt with immediately, then put a slot in your diary to deal with it later and move to the next email.

We must schedule our work as well as our meetings. If we all fall into the trap of just leaving our diary full of meetings we end up with just meetings and no room to do "real" work. I often hear someone saying they are "in back-to-back meetings all day" or "I am in so many meetings as can't get any work done".

It seems that this is exacerbated in the current increased remote working world we live in, though lessons are being learnt. If your diary is empty someone will put a meeting in, for goodness' sake when it is full, they will still put a meeting in. Protect yourself and your priorities.

I schedule focus work and use the Outlook quick actions to attach the email to a diary entry for each of the emails I am dealing with later. Then I file or delete them from my inbox.

Do It

Having evaluated the emails using the criteria above you should end up with emails that meets the following criteria

- It supports your business objectives

- No one in your team can deal with it instead of you

- It needs to be responded to now

- It will take less than 2 minutes to deal with

If it fails on any criteria, it should not be being done at this point and you should re-evaluate what you should do with it.

I Have Concerns....

What if my boss/customer/significant other sends me an email in between those 30 minutes?

I have a view which I imagine you will not be surprised by. If you do not schedule your email you may end up in back-to-back meetings, unable to check your email and not able to respond to your boss, customer or significant other.

By scheduling your review of your email, you can quickly see if they have sent an email and deal with it. By scheduling your review, you will have the time to deal with it. It may not be within milliseconds of you receiving the email, but it is likely to be that day or first thing the following day,

If it is critical, surely they will ring you if you have not responded?

Advanced Ideas

Once you have mastered scheduling and dealing with email in a focussed manner you may want to consider a couple of advanced ideas that I have toyed with, they did not work for me, but they may work for you.

Switch Off Email Sync.

By default, most email clients check your email servers every few seconds. We all know it is not immediate as we have all sat next to a colleague who is emailing something from three metres away and it takes ages to get to you. It must be, what one, maybe two minutes to get there.

There are settings within Outlook that allow you to stop it getting email unless you tell it to. Why on earth would I want to do that? I know that's scary, isn't it? What it will mean is that you have a barrier to checking your email between the convenience of clicking on your email client and just quickly checking before a meeting instead of building relationships with those around you.

Many of the tools are built for our convenience with a low bar for entry so everything is there in front of you even when you do not want it to be. By switching off the syncing you will no longer receive emails. WHAT! Hang on, let me finish. You will no longer receive emails until you decide to.

When you sit down to do your email, the first thing you do is press the send and receive button on your email client and then you'll get a bigger raft of items. Then you just get on and deal with them.

When I have done this, it has also shown me that I have a Pavlovian relationship with email. Once, I found myself between projects and therefore the email traffic was so low to the point I was convinced that the email servers were down, or I have a problem with Outlook. They were not and I did not. It was just that nobody wanted to write to me that afternoon, a bit of a blow to the ego, but I recovered.

Even if you don't feel comfortable doing this, I recommend switching it off when you are doing your email in those three slots, then you won't get the continued

email, email, email, which takes you on
beyond that half hour.

Set a Permanent Out of Office

When email was first about a colleague of
mine did this. I thought it was a weird
thing to do at the time for a few reasons.
Firstly, only dealing with email once a
day, there was not that many to deal with.
(Yes, it was that long ago). Secondly, you
are telling someone you are not going to
do anything about the email they just
sent and are going to ignore them until
later. Thirdly you would be sending an
email back for every email you received
filing up your email storage quota and
the sender's quota which was a big
constraint back then.

I have struggled with this one but for
some of you, it may work. My slots have
"Morning email moveable but ask first"
as the title in my calendar and I tell
people that I try to deal with email only
3 times a day.

Still Not Convinced?

If at this point you still have reservations about fully adopting this approach, then don't. I am not being petulant here, when I say don't I mean you don't have to adopt it fully. Try before you buy.

Try it for three days or a whole week and see how it goes. What is the worst that can happen? That feeling of dealing with a bunch of emails quickly will energise you and you will feel that efficiency. Be sure to give yourself a break too. If you miss a slot, so what, do it in the next slot. Or, and I can't believe I am saying this, if you check your email outside of the allocated slots, then so what, just try not to make it a habit.

This is all about improvement, small gains, not earth-shattering changes so try it and adapt it a little for you and your circumstances.

I would like to point out that I regularly fall off the wagon on this due to drifting back to my Unfocussed Procrastinator status. If you fall out of the habit altogether, then today is day one. Start again, and again, and again.

Outlook Rules!

(Other email clients are available)

Ahh sorry, not that Outlook Rules!
Outlook rules. OK let me start again,
Microsoft Outlook has a feature called
rules. This feature is incredibly
powerful and allows you to control what
happens when an email arrives or is sent
or several other state changes, for
example, if you flag an email for action,
it could change the colour.

As I mentioned earlier, I am not going to
give you blow-by-blow instructions on
how to configure the rules, check out the
details in the resources sheet, link in
the back of the book. I will however give
you some compelling examples of how they
can be used to increase your
productivity.

Carbon Copy Emails

I heard about rules many years ago on the
Manager Tools podcast and first used it
to deal with carbon copy (Cc) emails or
emails I have been copied on. For me, and
for colleagues I have shared this with,

it has been a game-changer in the ongoing fight to wrestle my email into submission.

With the advent of email, we can now copy the world. You will all have seen emails where the Cc is sent to sooooo many people that you're sure that somebody is worried about something. You may have been one of those copied in who don't care a jot about that particular email. What a waste of your time reading it.

Because these Cc emails arrive with no prioritisation or hint that they are Cc, we all appear to have really, really full inboxes. Those really, really full email inboxes can be a significant cause for anxiety, stress, inefficiency and ineffectiveness.

Just working out which emails you need to deal with is an uphill struggle. All you have to go on is the title, the name, sometimes whether it has an attachment or if the sender has put an important flag or action flag. Even the last two are only a reflection of the sender's view of importance and urgency, which may not align with your priorities.

This leaves you in a position where finding the wood from the trees is at worst almost impossible at best prone to error or misalignment with your priorities.

To help improve this situation I have implemented an Outlook rule which is triggered as each email arrives. If the email has me in the **Cc** box rather than the **To** box, then Outlook moves it to a sub-folder I have created below my main Inbox called Cc'd. You can do the same or call it what you like.

During my 30 min slot at the end of the day, if I have dealt with all the emails that were sent **TO** me, I have a quick glance at the Cc'd folder, though this is becoming more infrequent as the years go on. There is no point moving these emails and then checking the folder every five minutes instead of checking the ones that were sent **TO** you.

The first time I did this I wasn't convinced it would be of use and a little nervous that I would miss key emails but thought I would give it a try anyway. I set it up in the morning and let it run through the day holding myself back from checking that folder until the end of the day. As I said before I try to check only at the end of the day, though my lofty goal is to be confident enough to ignore them entirely, we can live in hope.

At the end of the day, I opened my Cc'd folder, I had resisted the temptation to peak, just. In the folder I found a list of about 20 emails all with the same subject. They ran through from early on in the day. Back and forth between different members of my team with many others copied into the email trail. Oh dear, I thought! (Or words to that effect, use your imagination.) It looks like something important has been going on and I missed it.

I delved in to have a look with my mind racing to come up with ideas to rescue the situation. I looked at the final email to see what the latest position was, to my surprise it said something along the lines of "That's great! All sorted, thanks for your help."

Hang on a minute, there's a bunch of stuff that's got done without me getting involved. How can that happen? I'm so important and pivotal to my team and organisation, somebody must have missed something. I had best get ready for a late-night fixing this.

I proceeded to check the email chain to see what it was about. Reading the first email, it was clear that had I not placed it in my Cc'd folder I am pretty sure I would have gotten involved in the conversation.

I read through to the conclusion of the discussion and to my disbelief I realised it probably got solved better than if I had been involved. My rule had worked exactly as I had intended, a Ronseal moment. It did exactly what it was supposed to. (For those who are not in the UK, the tag line for a decorating product company called Ronseal was "It does exactly what it says on the tin.")

If I had not placed it in my Cc'd folder, I would have been involved in resolving the issue. If I had not placed it in my Cc'd folder, I would have spent time doing something that when you look at it didn't need me. If I had not placed it in my Cc'd, folder I would have been diverted from my priorities. If I had not placed it in my Cc'd folder, I would have used the organisation's resources (me in this instance) to do something someone else could have done equally well.

Yes, my ego was bruised and just to polish my ego a little I convinced myself that I could have helped them do it quicker. Maybe I could have, maybe not. What is clear is that it was more productive to have me doing nothing in this instance than it would have been had I done something. (In the office I have noticed that productivity tends to improve when I am on leave. I have suggested that by keeping me away from the office, but with the looming threat of my return productivity would increase exponentially. None of my bosses have embraced the idea yet, but one can hope.)

By not getting involved in the email discussion it freed my time to deal with other things that had either higher priority or that needed me specifically and could not get solved without me. From that day forward I have kept that same rule running and I am convinced I

have saved countless hours over the
years.

FOMO (Fear Of Missing Out)

I mentioned this in the last chapter, and
I know for some of you the thought of not
checking, reading and dealing with
every email that is sent to you is
terrifying. I know some of you think
this is great news and will be shouting
"Woohoo! I don't have to read email ever
again."

It's somewhere in the middle. All of you
doing the Homer cheer, you still must
deal with email, it is part of your role.
No matter how ineffective or inefficient
we think it is, it is still the main
communication for most of us. So, buckle
down, use some of this advice and deal
with the emails that are left.

To address the concerns of the rest of you
who may be worried about such things as
emails getting missed or someone asking
you to do something in that copied
section, I have some suggestions that may
make you a little more comfortable.

Let's talk about the important client or boss scenario where they may have copied you on an email that they needed you to deal with. The Outlook rules tooling allows you to include exceptions to the rule, and I do that myself so if it's my boss, my boss's, boss's or my boss's, boss's, boss's boss or my..., well you get the point. I put their email address as an exception. If I receive an email from one of them that they've inadvertently copied me in on instead of sending it to me then it doesn't go into my Cc'd folder.

Hopefully, that addresses that concern of missing something from someone on your priority list. Whoever it is you can create different exceptions based on your needs. I'm not going to go into all the details on this. The resources link at the back of the book will help you set these up.

To address the second point where someone in your team or someone in the organisation just copies you and asks you to do something. I have a couple of points for you to think on. Number one, if it's important and you haven't responded or done it, or if you haven't acknowledged that you've taken that action, then it is incumbent on the other person to make sure that you have acknowledged you've got that action, not you. Unless you have spoken about it then they should not assume you will or can complete that action.

Secondly, to avoid them having the high ground, I recommend telling your colleagues how you deal with Cc emails. From then on it is not unreasonable to expect them to put you in the **To** box if they need you to do something. They will hopefully recall that if they need to ask you to do something they will put you in the **To** box.

Tag Your Mates

On the flipside, if we are honest, we have copied someone in and then asked them to do something in the body of the email, and failed to move them to the **To** box, we are not all paragons of virtue. To help me be better at this I use something I discovered recently.

When authoring an email if you place an @ symbol and start typing the person's name you can tag them. Same as social media and messaging channels. In Outlook I also found that if originally included the tagged person in the Cc box that when you tag them in the body of the email they moved to the To box.

You could think about coaching your colleagues on this behaviour as it will make it easier for you all in the long run.

Final Pitch

Just to try to finally convince you to embrace this I present a few experiences I have had.

Over the years I have mentioned that I do this to several colleagues and even customers, following my own guidance above to tell people. Whenever I've said it, I pretty much get the same kind of response, curiosity initially, a little bit of incredulity that I could be so blasé and then after a little conversation the question turns to how they could do it.

This repeated reaction is one of the reasons I feel it is worthwhile including this section.

I recently sat with a colleague to help them set up this rule for themselves after we had had the familiar conversation.

When finishing setting it up you get the option to run it on your bloated inbox there and then. So, we did! My colleague was delighted to see a couple of hundred emails disappearing from their inbox and reappearing to their Cc folder. That convinced her that it would work.

You Already Prioritise

Whether you think about it or not, you do prioritise your inbox, we all have to and because you sub-consciously prioritise, you do not read all of them. The thing is we are usually doing it with limited details or data such as:

- Who sent it?

- What the subject line is?

- Who else it is sent to?

- Does it have an attachment?

- Has someone set the high importance flag or the low one?

- Is it the latest one to arrive?

- You own random number generator

I would be incredibly surprised if any of these techniques align with your goals and objectives. None of these methods are effective. Even my method of choosing Cc's is not 100% effective. There may be information that you want in one of those Cc mails, so there's always a risk, but there's always a risk of without the Cc approach, just following any the methods above you may still miss that important information.

If there is risk anyway, why not own that risk rather than having no control over it and trusting to fate or karma.

Other Ways to Rule Your Inbox

In Lord of the Rings, there was one ring to rule them all, unfortunately, there is not one rule to rule your inbox.

But do not despair over your "Precious", you can create rules based on many criteria. The possibilities are almost endless so I will not be going into detail here. To decide what other kinds of rules you will benefit from, you will need to invest a little time thinking about the type of emails you receive and how you can start to deal with them differently.

I add new rules and remove old ones regularly. You may wonder why you should invest this time; you are busy enough aren't you? Let me tell you a story......... (Diddly do, diddly do, diddly do)

Imagine opening the email sent to the company bowling team mailing list reminding you to wear your new company bowling shirt for tonight's game. You were in the middle of your status report which should be sent in 10 minutes and now your flow is broken. You can't recall that key point you were just about to make about the project, so you must rifle through your notes again to remind you what they were.

Eventually, you find the info you needed but you just miss the cut-off and get a call from the boss asking to walk them through the project status because the report is not in the system. They only have time at 7.30 pm, just when the bowling game is due to start and to cap it all you had your shirt anyway.

Sound familiar? Maybe be a little exaggerated but I hope it makes the point. If you had set up a filter to move the bowling mailing list items to a separate folder which you reviewed at the end of each day, then you would not have been the cause of the company bowling team losing in the charity cup final, to the company's biggest competitor, resulting in them receiving all the free publicity for their community and charity efforts. Leading them to achieve greater sales and your company laying off people including you.... All from reading one low priority email. Sorry, I got a bit carried away there, but it is not unfeasible.

I have found many items that would fall into similar prioritisation. I use Outlook Rules to filter them out of my inbox. Here are a few ideas.

Training opportunities – an announcement from your training department or an external training organisation is not something that needs to be dealt with immediately or even on the day it arrives. You could put it into a folder that says training emails and diarise a slot where you deal with those emails.

Meeting invites – most meeting invites refer to the future not something in the next few minutes. Filter them out move them to another folder and again deal with them at a specific point in the day. On occasion, you do get extremely late notice cancellations or changes. Fortunately, these are few and far between and if you have checked your calendar before the meeting on Outlook there is usually something to indicate a change or cancellation. Worst-case, hopefully someone will message you if it needs to change. Again, it's incumbent on the person arranging it to be more effective.

All hands / Town Hall meetings – these kinds of emails are frequent and useful and can be important, but they are not usually time-critical. File them away, schedule a regular slot later when you have can focus on the meeting invites or output from the meetings.

Daily reports - Many of us have several types of daily reports that may give statistics on organisational performance, on your departmental performance or general information of the latest position on something. This kind of thing is information to do your job not to be dealt with immediately.

Admin stuff - Such as timesheet approvals, expense approvals, holiday approvals, both yours and your team's. Again, reserve some time in your diary to deal with them, whether it's every day, every week, whatever frequency you need.

It is more efficient to deal with stuff like this in batches. You log onto the system you are dealing with once and your brain is ready to deal with one thing, when doing timesheets your mindset is ready to think about timesheets. You will have the information prepared in advance too.

Important is a Lie

One final bit of advice on emails. If an email has an important flag, that does not necessarily mean it's important to **YOU**. Let me say that again it does not necessarily mean it is important to **YOU.**

It may be incredibly important to the person sending it, but it's your time and your priorities and your productivity that we are talking about. Don't assume that the important thing to them is important to you and therefore do not treat the email as if it's important to you, unless it is of course.

Do Not Disturb

As we discussed in the notification section, focus is critical to effectiveness. I am sure you have all heard of the terms flow state and focus work. There are many books written about how you perform better in this state and can be much more for efficient and effective.

We do have many distractions and entering this flow state is much more difficult than it used to be.

When you have a piece of work that needs your undivided attention then do all you can do to stop your attention from being divided.

We talked earlier about switching off your notifications and just focussing on emails, but I have started going further. I have started to switch off tools like Skype & Teams by setting my status to Do Not Disturb sometimes referred to as DND which I thought was Dungeons and Dragons, but hey ho. Try it and see what happens, see how much more productive you are. Even if people want to try to interrupt you, they can't. You are in control, as they said in Star Trek, "You have the conn (helm) number one."

When writing a document, switch off the phone, if you like to work to music, put on the headphones and go into your flow state. If your source of music is your phone, then switch off all other alerting features.

Instant messaging, whatever you use, Slack, Skype, WhatsApp, Teams etc. set that to DND and none of the messages will come through. Don't worry you won't miss them. They will appear in your email or in your missed messages. And even if you do miss them, so what. If it is important someone will try to get hold of you again.

If you can close your office door, then close it. Many of us do not have this anymore so use something to show your colleagues, or family in these days of working from home, that you are trying to get in your flow. Maybe draw the line at a big sign saying "F*&k Off".

The other option is to get out of your usual environment and don't tell anyone where you have gone. Climb a tree, climb a mountain whatever it takes to stop others from diverting you from your priorities.

Maybe remove yourself from the digital ocean we are all floating in and grab a pen and pad if the task at hand would allow it. Maybe invest in a Dictaphone for thinking time, we will talk a bit more about dictating in a later chapter.

Be Present at Home

DND is not only useful to be more productive in the office or with your work. Use it to be more present with your family and in your attempts to gain a solid work-life balance.

Recently I resorted to setting DND on IM during my lunch and in the hours after work. This helped me a lot when I had accidentally left the Skype notifications as it made sure I was not hearing them while having lunch. I was not diverted from my conversation with my wife thinking "I wonder what that message is".

When I did not have the DND on and some messages vied for my attention during lunch, I was neither in a state downtime, re-energising myself for the afternoon, nor focussed on the work items concerned. No one was getting the best of me.

At dinner with the family or significant other, put everything on DND, disconnect your watch and be with them as close to 100% as you can.

Mobile Subservience

Over the past 20 or so years we have all got used to having our smartphones and it is hard to imagine the world before we had them. I am sure my teenage daughter would be initially horrified if she were transported to a time before they existed.

Prior to the smartphone, we only had phones for making phone calls and occasionally text, that was all they could do, oh! and play Snake. Little by little they have evolved, I remember standing outside the pub on New Year's Eve trying to get a signal to text my brother a Happy New Year, now I ping a WhatsApp message to 30 odd people or a Facebook message to a few hundred in seconds, with a dancing gif of Carlton from the Fresh Prince of Bel Air. How things have changed.

With all the positives we have come to be dependent and as the title suggests above sometimes it feels as if we are there to serve these little electronic beasts not the other way round.

The first thing I suggest is to put the thing on silent. Not on vibrate, I mean on silent. This is based on a podcast I heard where they discussed a study on how devices create barriers to building relationships with others. It is an obvious barrier if you are sat opposite someone with your phone open, mindlessly scrolling. Most of us do it from time to time, it is not the exclusive behaviour of the teenager.

The study said even the act of placing the phone on to the desk next to you sends a subliminal message to the other person that you need that phone. The message you are sending is that a potential caller, social media post or email is more important than they are.

If the physical presence of the device is sending that message, what message is being sent when you have it making its little beeps. I have been in meetings when people have their phone set on vibrate so as not to disturb but because they have placed it on the table and for some reason, the phone company is using V8 engine to generate the vibration, off it went dancing across the table distracting the whole room. If you have not guessed that person was me.

I strive to not take my phone out of my pocket and revert to silent mode at work most of the time. I have opted to remove the vibrate too as every time it does vibrate, I think, "I wonder what that is, what if it is important?" Another cause of micro distractions, people can see when your concentration wanes, even for a few seconds.

To take this further, when having one to ones with people I have tried to not have my phone with me at all, leaving it in my jacket by my desk. Not on my desk ready to play my latest favourite ring tone to the whole office over and over or doing its vibration dance across the desk.

I take a notepad and pen, by doing this it immediately changes the impression that the others in the conversation have of your level of engagement with them.

You are not diverted by the phone's presence and your colleague is not subconsciously distracted by your phone's presence. The message you are sending is "This conversation is important; you have all my focus."

Taking a Break is Good For The Soul

Listening to another study, they talked about people becoming addicted to having their phones with them. Just a quick internet search will tell you that it is estimated that we touch our phones on average over 2500 times per day. That's an average with the 10% touching it over 5000 times per day. 5000!

Some people even get to the point when they do not have their phone with them, they become anxious and were even prescribed a block of plastic about the same size and weight to help ween them off their addiction.

I can easily believe this. I've found myself talking to someone on the phone, and this sounds ridiculous, looking around my house trying to work out where my phone was, I was a little worried and concerned and it took a while for it to sink in that I was using the damn thing. I am so used to having it in my pocket.

I try to detach my phone from me, but do not always succeed and get tempted back into the blue light glow promising delight. It is difficult but, in all situations, I urge you to think about what the priority is here.

- Should you be focused on this meeting to be able to contribute?

- Do you want to strengthen this relationship with this colleague?

- Or at home, sat at the dinner table do you want the family member to know that what went on in their day is as important to you as it is to them?

And as if to emphasize the situation I am sitting in my kitchen at home writing this. I am watching my eldest daughter with her mobile phone, headphones on juggling phone, milk and coffee cup and all manner of different things while still watching something on her phone. To be fair she's managing to do it all, but I suggest that she could be more

productive and quicker if she focused on
making the coffee and put the phone down.

How Does This Help
Productivity?

Aren't smartphones a boon to
productivity? Yes, they are brilliant
for improving productivity, especially
on the go. What we do with our
smartphones now is incredible. Some
people run vast business empires from
their phones. In some corners of the
globe, a mobile device is the primary way
to access the internet.

I am not saying they are to be discarded.
I run much of my life on mine and wrote
large chunks of this book on it. What I
am recommending is that you break the
chains when appropriate.

A critical part of a project managers
role is to communicate and build strong
relationships improves our ability to
communicate effectively.

Think about when you need your device
and when you don't need it. Prioritise
the person, not the tool.

Outlook Quick Actions

I've been a big fan of Outlook Quick Action since I discovered them a fair while back. I am not sure when I noticed them, I am sure no one told me. I have a feeling they had sat there at the centre of the Outlook menu ribbon trying to get my attention for some time. I am not always the most observant person around.

For those of you who are like me, they sit almost dead centre on the screen and have several different actions pre-programmed such as Move to a Folder or Email Team. The ones that got me interested were the Schedule Tasks and Schedule Appointment actions.

The quick actions are like macros in your inbox, and you can get highly creative with them. You select an email, then click on the quick action and it does what it has been told, automating all manner of day-to-day activities.

I have two main uses

- Scheduling work

- Filing stuff

Scheduling Work

I use my calendar to manage my larger pieces of work with smaller tasks on my to-do list. In both cases, I use the Quick Actions that allows me to schedule a meeting or task and attach the email in question to that task or calendar item.

Where I have a task that requires a little focus time, I use the Appointment Attach action which pops up a meeting invite with the email attached to it, I just need to choose the timeslot.

For shorter tasks, I use a couple of methods

- adding the email to my to-do list using a Task Attach action to create a task that also attaches the email and have a regular slot to review and allocate out time to deal with the email. schedule or

- using the Appointment Attach action for several emails to

bunch them together in a 30-minute slot.

In each of these methods that email is dealt with. You no longer need to use your memory or any other tool to remember to take action. I then either delete or file the email. Clearing down my inbox.

Filing Stuff

I also use quick actions to help me with my filing. I do not like to use my inbox to store every email I ever had. I have seen others do this and it would drive me crazy. I like to have some structure.

When I have any program or project to manage, I set up some folders within Outlook. The size and scale vary but I try to keep it about a max of 5 or 6. This is because I do not want a lengthy list of "quick" actions to scroll through as it defeats the object.

If you choose to go into several levels of granularity then I advise seeing where you drag and drop your emails to for a couple of weeks. Look at which are the most frequently used folders and set up a quick action to speed it up. As you can appreciate life changes so I review how I am using them regularly, though not particularly in a structured manner, and adjust to suit the latest popular folders.

Levelling Up

Hopefully, you can see how this can help you. Even if you have not taken my advice to schedule your email slots, if, as you are dealing with email you speed up your filing approach, that is going to increase your productivity, so look at them.

Like the rules, there are lots and lots of options you can use and even think about integrating the rule to trigger after quick actions. Now that is taking it to the next level, I have not yet got there, if you do, then please share your experience.

OneNote

The project manager daybook is ubiquitous across the industry. The mythical power it holds for each of us to be able to grab it and look back over the history of a project like some project management Dr Strange is critical for most of us.

Unfortunately, if you are like me, a handwritten daybook can sometimes not be as good as it could be. My default handwriting has in-built encryption which would make even the most talented cryptographers eyes water.

Even when I can read my chicken scratch words, deciphering the inherent meaning can be just as challenging, even to Alan Turing. Coupled with the usual back-to-back meetings we all have, getting a chance to write up and share the information can also present a logistical challenge.

I have experimented with tools such as Evernote and Word to capture notes but did not like having to type during meetings as it was a distraction to both me and the attendees. I continued to use my handwritten notes with limited success until back in 2015, when I was introduced to the Microsoft Surface Pro and its handwriting recognition, and the application called OneNote.

It was part of a project I managed the rollout of 25,000 of these devices. I got hold of one for myself to be able to understand the user experience (and I like a new toy.)

I had used a Palm IIIe several years ago, to do handwriting recognition and liked the concept, but the results and the need to learn a new alphabet reduced its efficiency. Once I got hold of a Surface Pen and OneNote, things changed.

Whilst it captured my handwriting almost exactly as pen and paper would, encryption included. It was also capable of deciphering my words too.

I know that the handwriting capability will not interest or be available to everyone but mention because I have found that this can help reduce the amount of rework of retyping up notes.

Killer App

I had described the use of OneNote on the Surface Pro as the killer app, for me it harnessed the power of both, but on its own OneNote has some fantastic features.

As I write, a basic version is bundled with Windows 10, the version I have used at the office comes with the Office 365 suite. Whilst using this version I discovered that it was well integrated into the Office product set and allows a multitude of hooks into each of the programs. From assigning tasks directly from OneNote into Outlook, to embedding office documents and annotating them, it became an invaluable tool.

Meeting Notes

The area for productivity improvement I want to highlight is around meeting notes. Microsoft has created the ability to take meeting notes that are contextual and linked to your Outlook calendar. The steps begin in your Calendar rather than in OneNote.

In your Outlook calendar when you right-click on a meeting you get a context menu. In that menu is an option called Meeting Notes. When you click on this you are given the option to take shared meeting notes or take notes on your own. As we all know the person with the pen owns the outcome of the meeting, so I usually choose the second one.

The first option can open up some wonderful team collaboration, as it allows everyone in the meeting to add to the notes. This can help share the note-taking burden but can lead to chaos if not managed well.

Having selected **Take your own notes,** you are asked where you want to store your OneNote page, then up pops a page for you to capture what is going on in the meeting, insert tables for actions and decisions etc.

The thing I liked was that it automatically copied the details from the meeting in Outlook. Attendees, agenda dates etc. and during online meetings I realised it automatically populated the in-attendance flag when people joined the call, which was cool.

With regular meetings I start the notes in advance of the meeting, copying across from the previous weeks notes the actions and any more detailed agenda items that have been suggested. I even use it for one to ones with my team and boss to allow me to capture things I want to talk about.

Easy Distribution

The final productivity thing I find of use is that I do not transfer the notes taken in this meeting to some crappy word template and then copy and paste who attended etc. I just use the button **Send as PDF.**

This creates a PDF version of the notes, automatically pops up an email with all of the invitee's email addresses pre-populated and the PDF attached.

For some reason, it is not in the default menus, so you have to hunt for it. I add it to the quick access menu by clicking the arrow and then navigating to **More**

Commands. Change **Choose commands from** to **All Commands.** Scroll down to the **S's** and double click on **Send to PDF** to add to your **Quick Access Toolbar.**

This chapter just scratches the surface of what OneNote can do; I have not yet embraced all its features but endeavour to do more with it.

Delegate

Before we start, what I am not going to provide here is information on how to conduct the act of delegating to your team. There is advice all over the Internet, my personal preference is the techniques suggested by the guys at www.manager-tools.com.

The key point I want to get across is to **DO** the delegation. Getting into that mindset of assessing every task and deciding whether you are the right person to do it or whether someone else in your team can do it.

It is a habit you need to get into and even if your technique of delegation is not that great and you have to redelegate a couple of times, the fact that you are assessing your work and making priority calls on your time will improve you and your organisation's efficiency and longer-term the effectiveness of you and your team.

Maybe it will give you some extra time to work on your delegation techniques.

Why Bother

Delegation is one of the most useful productivity tools we have at our disposal. Both individually and organisationally.

Individually it means we can free up some time to work on the priorities that are important to us, be they career or personally driven.

Organisationally it means you are working on the organisation's priorities as well as training the less experienced members of your team. The organisation gets two birds with one stone, more skilled staff and more focussed management, what's not to like.

The thing is most of us have a way to go to improve delegation. The effort in delegating and getting the job done is not insignificant at first. Here are some of the excuses, yes, they are excuses, that I have used or thought of. You recognise a few:

- By the time I have shown them how to do it, I might as well have done it myself.

- They are busy with their stuff; I will just cover it off.

- If I don't do it people will think I am passing the buck on my responsibilities.

- They won't do it as well I can do it.

- I need to get approval from my manager to do it.

- I need HR to approve the job role change.

- I need to update the RACI / job description first.

- I need it now. I can't wait for them to do it.

- But I like creating fancy Excel spreadsheets.

Recognise any or all of them? As managers, we are stewards of the organisation's resources. Resources also includes the squishy human resources.

We would not tolerate wasting organisational cash on things that could be done at a lower cost, within tolerable time and quality measures, would we?

Why then, do we accept it in the allocation of our own time. I heard a quote on the manager-tools.com podcast describing it as management 101. It is the responsibility of every manager to ensure, that if a task needs to be done, then it should be done by the member of the team who influences the smallest part of the organisation's resources, providing they can do it to the right quality and in time.

Now there are exceptions where we need to throw all hands to the pumps and have spent time handing out laptops and setting up machines during rollouts. These exceptions will always be there. What we are talking about is the normal day to day operations.

Just a quick point, I am NOT advocating delegating all the things you do not like to more junior staff. That is just being, well I can't use the words here, you get the gist.

The reason organisations exist is to provide leverage. Each of us independently trying to get something done is hard. It can be done, but it is difficult. As soon as you have someone else with you your combined ability more than doubles. You have the other persons experience and ideas to join with yours, creating connections and ideas that would never come to you on your own.

As you increase the numbers of people you can influence, the act of not delegating becomes more expensive. If you spend one hour pulling together data for a report and making a flashy Excel spreadsheet, it's a waste. If you have a team of 50, several members of your team could probably have done that report, with 15 minutes of direction from you.

It costs you 15 minutes, it may take them two hours to complete, they may not be as much of a wiz on Excel as you. The report has cost the organisation two hours fifteen. Is it better or worse than you doing it in an hour?

The answer to that is; it depends.

- It depends on what you do with the 45 minutes you saved.

- It depends on whether you make it part of that person's job.

- It depends if their need to learn a bit more advanced Excel skills increases their capability and allows them to take on a few other items from you.

It depends if you look beyond the task at hand, lif your head and look at the wider picture.

If with the 45 minutes you spent, you could pull together an email gaining approval for expenditure to save the whole team fifteen minutes a week by introducing some new software or removing some inefficiencies from processes. That means a saving of 750 a week for your team, 12 and a half hours a week or about 75 days a year. For the cost of two hours and 15 minutes.

It's a no brainer.

Lift People UP

I saw a comment on LinkedIn

"...pulling people up,

not standing on them to succeed"

Gary Vaynerchuk

Delegation enables us to do this "pulling people up." More than that, as a manager you should be thinking about succession planning for your role. Succession planning internally is cheaper and better than recruiting externally so it's part of your job to effectively build your potential replacement, should you move on.

You can send people on hundreds of training courses, but they still miss out on the feeling of doing the job.

Many studies talk about learning and break them down into

- Hearing

- Seeing

- Doing

- Teaching

The role of delegation is to allow someone to apply the skills they have heard and seen in training courses and try them out in the real world and subtly adapt them to their style.

When you teach someone on your team you have to look at what you are doing more critically to enable you to explain it well. This will inevitably lead to you finding tweaks to improve what you do on this task and other areas of your work that you decide to retain.

An Example

Let's take a simple example of your monthly project board meeting. Say it has moved to a date when you have your holiday booked. It's 3 - 4 months away, but you cannot change the holiday. I have seen several approaches:

- Try to cover your input offline.
  - No chance for discussion or elaboration.

- Attend remotely from the beach – Hassle for the whole of the holiday, no chance to properly relax. Resentment from the family that you are distracted leading up to the meeting. Worrying all week that your precious company laptop, that you have gone through levels and levels of authority to take it on holiday, may get stolen from the hotel safe. When you do attend, the hotel Wi-Fi is being hammered by all the Instagram and Twitter posts of people's holidays and no one can understand you on the call anyway.

- Your boss attends in your place – They are not up to date with all the detail. They may not have the same relationships you have and when you come back you have a bunch of things agreed to that you were trying to avoid. It also costs more and takes your boss away from their leveraged work for the organisation.

And Now For Something Completely Different

How about something different. Get your most appropriate team member. Coach them over the next couple of months in managing the project boards. Let them run one beforehand or provide your contribution with you in the room to assist. It develops them, and if they do well, you could continue the situation freeing up your time to focus on the more subtle elements of the meeting or not attend at all. Most important of all you can lay back with your Mojito and relax by the pool.

Choosing What to Delegate

Now I have convinced you to delegate, you have to choose what to delegate.

Having a solid view of your priorities is key to successful delegation. If you do not know what is important how can you decide what are the less important items that you can hand to others?

Sometimes we think we are the only ones who can complete a task, or we enjoy it.

These tend to be the hardest to let go of.
If you have a clear view of your
priorities and honestly evaluate each
task against these priorities, then you
will have to let go of them. Use these 4
steps to decide if you should do it.

- Does this task support my
  priorities?

    o Yes - move to the next
      step.

    o No - delegate to one of the
      team.

- Do other people in the team have
  the skills to do it?

    o Yes - delegate to one of the
      team.

    o No - move to the next step.

- Could you coach one of the team
  to do it?

    o No - next step.

- o   Yes – coach one of the
    team and
    delegate.

- Get the most appropriate team
  member to identify a coach for
  this task, ready for next time
  and get them to shadow you while
  you complete it this time.

I know this eventually leads to you
doing nothing. Not ideal, but then again
if you are looking to take on a bigger
team, larger department or projects, it
gives you headroom to do that. As ever
there will be some things that fall out
the bottom and you will do them. That's
great. They will be the most important
things that only you can do.

I hope this has convinced you that
delegation should be part of your
toolkit.

Prioritise

Why Make Priorities a Priority?

Throughout this book, I have given you hacks on ways to execute on your priorities. What I have not covered is deciding what your priorities are.

Over the 30 years of my career, I've attended many different time management courses, general management courses, read literature on the topic and listened to podcasts on being a better manager.

It should be no surprise that knowing your priorities is always referred to as being a massive influence on whether you are effective in your chosen role.

I decided not to tackle this topic at the beginning as I did not want you to get bogged down in establishing your priorities before you start being more efficient. The reason for this is that working out what your priorities are is difficult and time-consuming.

If you already have the need to make your days more efficient then carving out the time to carefully consider your priorities could have been a barrier to entry. By implementing some of the suggestions in the previous chapter you may have become more efficient, now let's crank it up and help you be more effective.

If you do not know your priorities, how can you effectively make decisions on your workload? When I have not clearly understood my priorities and consciously evaluated tasks against them, I have quite often found myself allowing other people to decide what I do next.

This is what we end up doing by dealing with the next email in our inbox or going to the next meeting at someone else's request assuming you must be there.

You are not deciding what to do next. You are handing over all the control, the results and ultimately the success of your career to others. Now, that's fine if you want to do that and have an "easy" life and not rock the boat. Remember an easy life may not be a pleasant one.

I think of it differently, when my daughters say "I don't mind" about something I quote the lyrics of one of my favourite Rush songs Freewill:

> **"If you choose not to decide,**
>
> **you still have made a choice"**
>
> *("Freewill Lyrics | Rush.com")*

I am usually greeted with rolling eyes. I like the lyric later in the chorus

> **"I will choose a path that's clear,**
>
> **I will choose Freewill"**
>
> *("Rush - Freewill Lyrics | AZLyrics.com")*

Without knowing your priorities how could your path be clear? It won't be, will it? You will be choosing not to choose.

The key point here, like many of my points, it is not the tool you use to prioritise that matters. It is the action of identifying and ranking your personal priorities that matters. There are a multitude of tools and techniques to help you work out your priorities and each of them has merit, some of them you will like, some of you will not.

I recommend trying some out and seeing what happens. Remember it is not a one and done decision, you can try something for a week and if it does not work for you, try another. Even if you sit and write down your top three, five, or ten things that are important to you and rank them 1 to 10. You can take each task and see how they measure up in achieving these priorities. Even this simple approach gives you better data than none.

Priorities Help Time Management

I know from bitter experience that to have a productive approach to time management, you must have a view of your priorities. Every task we have in the office (and at home) has an inherent priority associated with it whether you consciously or subconsciously allocate one. Sometimes the priority that is subconsciously allocated is not reflective of your conscious priorities because we have not made the conscious assessment of the task against our priorities.

Storytime

The following story illustrates how if we prioritise our priority tasks first, we can complete more work.

Through the doors of a college lecture theatre, a professor enters carrying a large glass jar which he places on a table next to the lectern. Alongside he places his bottle of water, five big rocks, a bag of pebbles, a bag of gravel and a bag of sand.

Placing his folder on the lectern, he turns a few pages, looks up and claps his hands together. Waiting for the murmur of conversation to subside and without saying a word, he starts placing each of the rocks into the jar filling it just up to the top.

"Who thinks this jar is full? Raise your hand." all of the class who were listening raised their hand.

Without a word, he picks up the bag of pebbles and pours them into the jar.

"Who thinks it is really full now?" most of them raised their hand but a few were considering the remaining bags on the desk and kept them down.

Unsurprisingly he turned his attention to the bag of gravel and poured it into the jar.

"What about now?"

The class were now wary but most of them said it was full.

Finally, he poured the sand in and filled it to the top.

"Full this time?" he asked. The whole class raised their hands convinced they were right this time.

Picking up his water bottle from the desk he cast his eyes around the students, looking each of them in the eye as he took a swig from it. His thirst quenched, he upturned the remaining water into the jar, emptying the contents.

This story is used to illustrate many things and the origin is fairly foggy.

When applying to prioritising your work, you will have some big important and complicated tasks, the rocks, some less important or less complex ones, your pebbles, and so on down to items like the water.

If you fill your jar first with your low importance and low complexity tasks it will appear full and it will seem that you cannot fit in anything else.

This is what it is like when we let email, IMs, filing and other low priority tasks take the lead in our day-to-day work. It feels like we are busy but never get time to do the things we need to do which service our main priorities.

By scheduling your high priorities items first or dealing with the first thing in the day, you are putting your rocks in your jar, and you can see easily that you have room to add in your pebble tasks.

If you throw in the water, then the sand, then the pebbles, the jar is too full to put your rocks in. Then you end up getting a bigger jar by working late, working at home or failing to deliver on your priorities.

Without knowing and being able to assess each task you have, against your priorities, you will not be able to make these decisions. Which task is a pebble, which is water, which is one of your rocks?

Goals and Objectives Make Prioritising Much Easier

To work out your priorities is not a five-minute job. It will require effort. It all starts with you knowing your goals and objectives. Otherwise, you are prioritising based on someone else's goals and objectives. Whilst this is still better than not doing it at all it may not be as fruitful from a personal perspective.

Do some research, do some soul searching for your personal life and your work life. What are **YOUR** priorities? Not your organisation's, **YOURS.**

If you choose a goal that you want to get to the top of the tree in your organisation you may need to reduce activities at home or leisure to accommodate this. If your chosen goal is to have more time at home than you do now, then you will need to think about your priorities in the context of getting out of the door and switching off from the job.

I am not advocating either of these, the choice is yours and yours alone. If you have goals, then make sure that whatever you are doing each day supports these goals.

Project Priorities

Project managers, frequently have a massive list of things that need to get done as an individual and as a project team. Dealing with individual tasks and assessing them against our own priorities is sometimes simpler than assessing those of the project.

All of the tasks have a number one priority in the eyes of different stakeholders. This is obviously untenable, if we tell our project team that every task is the number one priority, then nothing will get done, the team and the project will be stifled.

If everything is a top priority, then nothing becomes a priority.

Ultimately, our priorities tend to lean towards the 3 pillars of time, cost and quality. Whilst there are subtleties underneath this one of the pillars tends to have the highest concern for the organisation. I recommend reviewing your project and deciding what is important to your organisation.

Is it driven by?

- Time such as legislation or to react to a global pandemic?

- Cost, such as a government-funded project with no option to obtain more?

- Quality or scope, such as a commercial flight to space,

where making it halfway or with only some of the passengers returning is not an option?

I do not suggest throwing the baby out with the bathwater and just focussing on one goal or pillar. If the main priority is time, then you need to assign a higher priority to tasks that relate to achieving a specific date, perhaps delivering some things in a future release or requesting more budget.

When you have established these priorities, you must share them with your team. They will then not be diverted onto low priority items. Tell your stakeholders too so they all understand your decisions. They may not be happy, but they will better understand what you are delivering and why.

Doing it

It is way easier, from a short-term basis, to not do this rather than to do it. I struggle with this, and you may too. It takes discipline and consistency. You will occasionally fail at it; you will let it slip a little at a time and then realise suddenly that you are no longer doing it.

That's OK. Give yourself a break, and at the risk of repeating some of my introduction, just like any changes in your life, today is the first day. You can't change what you have done, there is no point lamenting your past failures. Just start again and make it a priority.

Remote Working

I couldn't write a book about productivity during 2020 and 2021 without talking about remote working. Whilst relevant before it the COVID-19 pandemic we have all experienced the rapid acceleration in the adoption of remote working.

Before I start, I have no medical training and all the things I mention in this chapter relate to my experiences with mental and physical health and a few things that have worked for me. Please obtain professional medical advice if you have similar issues, I have done, but your circumstances may differ from mine, so make sure you get advice from an expert.

My Experience

Since early March 2020, I began working at home five days a week rather than the usual once a week with the occasional foray into two days a week.

With many of us moving to this remote approach overnight, just about every conversation ended up being Skype, Teams or Zoom call.

The level of usage went up through the roof because it was the only way to communicate. Suddenly I found that my days would be full 30-minute back-to-back calls.

Previously I would do one day where I did the pick-up from the school run, but little by little I missed these days due to constant meetings.

As well as the full-on days, I noticed that we were no longer doing our daily commute and these calls were starting much earlier and running through until much later than they normally would do.

Dungeons

I am extremely fortunate, that I had a separate room to do my work. I did not have to camp or battle for space in the rest of the house. The room is however quite small, and technically underground, with no natural light and a desk that faces a wall.

Over time I started referring to it as my dungeon. I seemed to go in there in the morning and then hardly leave it for the day. Although initially a joke, in hindsight this was not a very constructive term to use to maintain a healthy mental perspective on that location.

It used to be a novel place to work from home, and when doing other hobbies like writing or podcasting it was my little bolthole where I could get on and do some stuff. As it was unusual for me to be in there I tended to be left alone by the family.

During lockdown, my relationship with it changed. It became a place that I would try not to be when not working. When I was trying to do something that wasn't related specifically to my job, I found it much harder to get it done in there.

The non-optimal setup at home and sitting at a desk constantly wearing a headset limiting my movement created a recurrence of some shoulder problems to the point that I could only be comfortable lying on the floor. I even attended video calls lying down on the grass outside as it was the only place I was not in pain.

I eventually figured out was that it was not necessarily the actual setup I had, but more the lack of physical variation in my body position during the day. This impacted my productivity, even leading to me being off sick for a day or so.

Decompression Depression

As I no longer had to commute, I lost the two fire breaks I had between the office and home. During my forty-minute drive, listening to a podcast or some music, I would have the chance to mull over the events of the day on the way home, ponder problems that needed sorting when I arrived, solutions to which quite often came to me in one of those commutes.

I would get a short walk from the car to the office. I noticed that I was not getting outside as much, even when we were able to. My daily steps were really low ☺.

I had to leave at a certain time and get on the road to the office or back home, so the early starts and overrunning calls at the end of day, which sometimes overran by hours, were not a frequent feature of my day.

Frying Pan to the Fire

When I had finished my day at work, I would step out of my office into our dining room. Initially, this was OK, I thought it was great, not being stuck in traffic and missing out on spending the time with my girls. But over time I found a couple of things.

Firstly, as the days were longer at both ends, the time saved from my commute was being flittered away. Rather than being able to step out of the office to either help with or sit down to dinner I was more frequently stepping out partway through the meal.

Secondly, I started to miss the decompression I had in my commute. Stepping out of my office straight into the dining room and straight into husband and dad mode. No opportunity for me to process the events of the day.

With no chance to decompress and coupled with the poor ventilation, lack of light and the downturn in daily movement physical activity I would gradually find that through the week I was getting more and more exhausted. Having eaten dinner and sitting watching TV I was just flaking out fast asleep. Something that I would rarely do before.

What Did I Do?

Before telling you what I did I want to mention that I first started writing this section when we were deep in the lockdown and am now editing just as we are leaving it. The things I have put in place I will continue to use when working remotely as they are still relevant but maybe not as extreme in causality as it has been.

Take a Break

Take physical breaks. There is a tendency to feel that you need to be always-on when you are not physically in the workplace. I have felt like this, especially with the rise of instant messaging.

To break (no pun intended) this, I challenged myself by asking a few questions:

- Did I expect others to be "always-on"?

- Did I expect others not to grab a coffee?

- Did I expect others not to have their lunch?

- Did I expect others to not go to the loo?

The answer was, unsurprisingly, no! My expectation of others was the opposite, I encouraged them to take these breaks. So why was I holding myself to a different standard? I have not got a great answer to that and to be fair it is pretty irrelevant why I did it and if you are doing the same the reasoning is irrelevant too.

What is relevant is the identification of that behaviour and taking action to redress it. I took a couple of actions and they related to my Outlook Calendar.

Shorter Meetings

I set up meeting or work slots in my diary to be shorter than the default hour and half hour and run meetings to finish at these shorter times.

I discovered that Outlook introduced a feature that allowed you to change the default duration for calendar meetings to be 5 minutes shorter per half-hour. This automatically builds in time between the end of one meeting and the start of the next.

We have all experienced the back-to-back meeting syndrome where as soon as one of the sessions overruns the knock-on effect is that you are late for every meeting.

When you add the remote nature, you tend to be less later, as you are not walking from meeting room to meeting room, but the effect is this exhaustion of not getting the break.

Remember Parkinson's Law from earlier, the same applies here. Most meetings and conversations have a level of non-critical content or repetition that, if left unchecked, will expand to fill the available time. Especially if one of the attendees loves the sound of their own voice.

If you chair the session, then you can influence keeping to time. Even if you are not, you can assist the chair by being concise yourself and discouraging the rabbit hole conversations by saying to your colleagues "Can we pick this up outside the meeting? "

When it is a scheduled task, you have absolute control over getting it done in the scheduled time.

If you manage to finish half of your hour-long meetings in 50 minutes, it gives you 40 minutes a day (assuming an 8-hour day) to pop to the loo or grab a drink or just walk and stretch.

This applies not only to remote meetings but face-to-face ones too.

Schedule Breaks

Schedule breaks, lunch, coffee, etc. and state "DO NOT BOOK" in bold on your calendar. Reject any invites that clash.

Simple as that. Oh, one more thing, take those breaks, don't just use them to catch up on email or IM. Take the breaks away from your desk, work phones and laptops. Guard them as if it was a meeting with a very important client or your boss about to give you a pay rise.

If **YOU** don't, no one will. Other people won't know how important they are and will override them for their priorities.

You must be ruthless because these breaks will refill your energy well. On days when I have been strict, taking a full lunch and breaks my effectiveness and my productivity in the afternoon is increased noticeably.

There is a need to allow some level of flexibility in order to effectively support our organisations, but be incredibly careful, as one chink in the armour can lead to you allowing everyone to override those sessions. Guard them, but don't be belligerent.

Zoom Society

During these last 18 months, we have evolved into a video call first society. The tools have been available for ages, both professionally and personally. Dare I say the younger parts of society seemed more comfortable collaborating over video, watching the latest Netflix film whilst on a video call with each other. Some of the older members of society taking a bit longer to embrace it.

With lockdown suddenly everyone started using video to keep that face-to-face contact. Which was great, to begin with.

Over the months the energy required to be on screen seemed much greater. I am sure some experts have been studying and have a scientific explanation why I started to feel more exhausted when on video for long periods.

I liken it to being the presenter in a meeting. When you are a participant in a face-to-face meeting, the focus is not on you, it is on the person talking at that time. Not everyone is looking at you. That's why you surreptitiously check your emails on your phone occasionally when you are not interested in part of the conversation. When you are running a workshop or series of meeting you find that at the end of the day you are shattered.

When we are on video, we can see everyone on your screen. And everyone can see you on their screen. Even though we may set the focus to be on whoever is speaking we still feel as if we are in the limelight. If we look away to another screen and not at the camera, we feel people may think we are doing email or watching the TV. There is more pressure and perceived attention.

Considerate Video

Prior to the pandemic I would be a "video on first" person and encouraged people to use it. This was when we were doing less frequent remote calls, still having those face-to-face meetings too.

After a few months in I changed from video being the default to being the considered decision. I use it when I want to connect to build a better relationship. Whether it be a new personal meeting, a one to one with one of my team where I want to be able to better assess how they are doing or with my boss or customer covering an important conversation.

I also take into consideration the person or people I am talking to. Depending on their character, or their set-up at home they may feel uncomfortable with video being on. If you are cajoling your team into using video, making them feel that they should have it on, when they prefer not to, you will reduce the effectiveness of that interaction. They may be self-conscious and have difficulties with the cameras on. Maybe have a Zoom background of the week challenge for the whole team. You could offer a small prize for the best one as voted for by the team.

Ask

The level of success that has been witnessed with people working remotely will inevitably result in us all working remotely or working with remote team members regularly. The only way you will know if they have an issue with video is if you talk to them. Ask if they have problems with having the video on.

If they do, ask why, you can then make allowances or help them get around the problems, maybe coach them on using the backgrounds features on the video apps to hide their pile of ironing. On the flip side, if you have some preference or constraint tell them.

By reducing your on-screen time, you will notice a reduction in that performance fatigue.

Love Your Musculo Skeleton

When sitting in the office we have a lot of support checking that we are correctly seated and have the right set-up. With the move to more remote working the risks to our posture and musculoskeletal health have changed.

To resolve my physical issues, I conducted a few experiments:

- Speakerphone - for all my Teams calls, I changed from video to voice and set the laptop to speakerphone. I had the advantage of having a door I could close. I would then stand and move around my office rather than being stuck to the chair and headset.

- Bluetooth headset - this allowed me more freedom of movement and security as only I could hear the other end of the conversation. The only thing I found was that my ears were sore having an earpiece in all the time which led me to mix it up a bit.

- Walking Meetings - I started looking at opportunities to walk when on a call. Considering who could overhear and whether it was confidential or if content was being presented.

- Moving about the house - getting away from my normal seating position. I would move the laptop to another part of my house or if the weather and screen allow, I got out into the garden.

- The Pen is as Mighty as the Laptop - technology is great, but overuse can be a problem. I went a bit retro and picked up a pen and pad and moved away from the machine I needed to retype my notes, but it gave me a break, mentally and physically.

Get The Right Setup

All of us have a different situation at home. Some of us have a dedicated room that we can use for doing our work. Some of us may have a state-of-the-art garden office. Some of us are balanced on the spare Christmas chairs on a table in the corner of the living room. If your set-up

is poor then think about it, short-term
your body will be OK but long term it
will lead to health issues.

If your organisation provides
equipment, then jump at the chance. If
not, you may need to invest yourself. Get
a good chair, appropriate size desk,
monitor, keyboard etc. Don't be hunched
over that laptop on your bed it's not
sustainable for long.

Set Boundaries

Working at home means that you are
encroaching on your personal life. If
your setup takes over the whole dining
room this will create tension.

Agree with your significant other on
what you are going to do. If your only
option is the dining room, then agree
what time you will be using it and clear
the space when you have finished.

Once I was on a call and my door was open.
My children came back from school and
one of them was making loud noises, as
kids do. My wife, realising I was on a
call, asked her to "Shush". Overhearing

this I felt guilty. It's **THEIR** home, and my work should not drive what happens in our home.

I should have had my door closed. When I came off the call, I apologised and said to them all that it was not up to them to be quiet when they get back from school, it was up to me to make sure my door was closed.

Set those boundaries, tell them you are going to be working, and agree where in the house you will be and if you need some quiet ask for it. Don't just expect it.

Remember it will not be perfect and everyone who is on a call with you has the same issues working from home. Very few, if any will care about a bit of family background noise.

Face to Face Still Exists

Yes, you heard me. Remote working is here to stay but we will still see each other in person.

I am quite comfortable working at home for the foreseeable future. I plan, when it's practical, to start putting in place face-to-face conversations with team members, customers, stakeholders and my management.

Whilst I can build relationships over the phone or video call, none of them will be as effective as a face-to-face conversation.

Many years ago, I worked with an offshore delivery team and had two main contacts who took our requirements and passed them to the developers. Initially, we had some difficulties with communication which led to misunderstandings from both sides of the team.

To resolve this, at not an inconsiderable expense we flew the two of them over to the UK. They worked with the UK team and the customer for six weeks, they built good relationships and an understanding of the customers business which over video and phone calls had proven difficult to convey.

The quality of work and reduction in misunderstanding were noticeably different and the dynamic between the two teams transformed to that of one team rather than two separate organisations.

Most of us will move to a hybrid model and when you are in the office don't just sit on email or conference calls, sit with people, have a coffee, work something out on the whiteboard together. Do the things you can **ONLY** do face-to-face. The other stuff you can do when you are back home in your pyjamas

If you are someone who would prefer to be at home and not have to meet others or like the fact you do not need to commute, I urge you to build in these face-to-face slots.

Be a Dictator

No, I am not recommending that to improve your productivity you conduct a military coup and take control of your country... though having said that, you may be able to get some things done a bit easier. (No. Really, don't do that!)

This book was drafted using the inbuilt dictation capability in Microsoft Word on both the desktop version and the mobile phone app. Before the advent of computers, a considerable proportion of professional writing was completed using dictation. The originator would either speak to the person who would do the typing or record it on a cassette which was passed to the typing pool or secretary to listen back and type up.

With the introduction of word processing and computers, the typing creation moved to the originator of the documents. For some, that is not too bad as we may be able to type quickly but I tend to find I write differently from how I speak, and more effort is required.

When authoring books, the benefits are obvious because the pace at which you can get a first draft written is way above that of typing even if you're a quick typist.

Now don't get me wrong the technology is not perfect, but neither is my typing. (Please send all typos spotted in this book to typoalert@nigelcreaser.com) Sometimes it mishears me because I mumble a little or my enunciation needs to be improved. Some of the errors are also very amusing to read back, generally the rude ones.

The speed at which I was able to create this draft was way above any other books. What I am finding is that the editing is a bit more time-consuming.

The wonderful thing with a mobile app is that you can also wander around dictating on your daily walk or roaming around your house or wherever you want to do it. This helps you break away from the desk. It allows you to do two things at once, one which does not require your attention (walking), so it still meets my call for us to focus on tasks and take care of our posture.

You may be thinking how is this relevant to me, I am not planning to write a book. I use the book as an illustration and the experiment I have been conducting. At the end of the day, this book is just another document. As project managers, we generate a vast quantity of documentation and anything that can help us speed this up will be of benefit. So, considering your organisational security considerations you can use dictation to produce proposals, business cases, status reports, emails, even instant messages.

There is a multitude of ways to dictate and here is a list of ones I have used, which at the time of writing are all available:

- Dragon Naturally Speaking – this was around back in the '80s as is an extremely popular piece of software, I have not used this in recent years, but it seems to be the premium offering.

- Microsoft Dictate - part of the office product set, simple to use and reasonably accurate.

- Otter.ai - mobile phone app with something like 600 minutes free or you can pay for more. You can also upload audio files and it will transcribe them.

- Voice recorder on your phone and upload to Google Speech to Text or Amazon Transcribe.

- Create an audio file and send it to a human transcription service.

As I mentioned editing can take more effort, so be prepared to have a little more time spent fixing some of the errors. My experience is that I still get the benefit of getting the raw document down outweighs the editing time.

Getting Over Vocal Hurdles

When using these tools it takes a little while to get used to you recording your

thoughts. Here are a few thoughts on
helping get into it.

- Don't worry about document
  structure - you can always move
  it around later.

- Do not watch the computer type -
  I found it distracting as I could
  see when it misheard me and I
  found it hard to keep myself
  from stopping and editing at
  that point, defeating the object.

- Leave yourself notes - you may
  need to research something or
  ask someone something, just say
  something like Editing note
  blah, blah, blah.

- Go over the audio - you can
  always go back and listen to the
  actual audio later if the tool
  you use records it.

Giving the Dictator Backchat

A related hack here is that in most of the
Microsoft product set and similarly in
other office applications there is a

feature called Read Aloud. This is a text to voice capability. It will read out what you have written.

It is fascinating to find that when a sentence you have written and re-read and re-written several times is read aloud by someone or something else you will find a whole host of errors. Having read something many times our brains seem to go "Don't worry I know you have loads of things wrong, but I know what you mean" and it fills in the gaps or fixes the errors.

I read somewhere that if a word has the first and last letter correct you don't need the letters in the middle in the right order as your brain will shuffle them around for you. Clever, but not useful when you want to produce a quality communication product.

Despite, or maybe because of the robotic sounding voice, you can pick up where the flow of a sentence sounds wrong much easier than looking at it. The gaps are not plugged and typos un-typoed or de-typoed by the machine, it just reads what you wrote.

Thank You

That's it I have finished with all the hacks that I wanted to share. There are a few others that I have tried but I am not convinced they are fully elaborated yet so will keep them close to my chest for now. As you may have noticed this book is the edition for 2021 and I plan to add and enhance it over time so keep an eye out for more hacks.

Hopefully, some of these hacks, if not all of them, will prove useful to you and again I reiterate what I said at the beginning of the book don't try to do all of them at once. Your head will explode! Whenever you try something new, there is a time investment to get it started, pick one that resonates with you or just the first one and use the time saved from that one to instigate the next, and on, and on.

I know that if you implement just one idea from this book you will gain back time for you to spend how you want.

Throughout the book, I've mentioned several different tools and resources that will help you improve your productivity, rather than have a constantly evolving list in this book that will require updating I have put together a list on my website which I can update more easily so pop along to:

www.nigelcreaser.com/pmprodhacks.

If you enjoyed the book, please jump on to one of the online bookstores and give me a rating and a review, especially if it's a good one.

If one of the hacks has helped you out, I would be delighted to hear about what it was and what you did with the time saved. Drop me a message on one of the these:

hacksdone@nigelcreaser.com

www.twitter.com/sundaylunchpm

www.facebook.com/sundaylunchpm

www.instagram.com/sundaylunchpm

www.linkedin.com/in/nigelcreaser/

Thanks for spending your hard-earned cash and good luck with your productivity journey.

All the best,

(See I told you it was encrypted.)

Dedication

To my beautiful wife and daughters who have no idea what I do for a living. They have as much knowledge as the characters in Friends do about Chandler's job. So I just say I am a transponster.

Also By

Nigel Creaser

When I Were a Project Manager

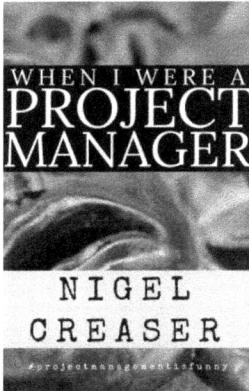

Where it all began...... Are you a new project manager wondering what life might be like at the end of your career, maybe reminiscing about the journey with colleagues? Do you want a tiny glimpse into your future?

Are you an accomplished project manager who recognises the funny stuff in the life of a project manager?

Well, this is the book for you! A parody of the "Four Yorkshiremen" sketch adapted for the project management community.

Peter Taylor, author of The Lazy Project Manager says:

> *"Python meets Project Manager! As a Monty Python fan and Lazy Project Manager, I just loved this excellent reworking."*

Pick it up here <u>amzn.to/2Xyjhvr</u>.

Project Management:

The Sketches

This book has been described in many ways,

"...I laughed so hard I nearly peed my pants."

"This book single-handedly help me pass my PMP, APM, Prince 2 and grade 3 bassoon exams."

"...the funniest book on Project Management I have ever read."

"Well, it made me laugh."

None of these statements have been said about it, except the first one, Peter Taylor, The Lazy Project Manager said that and he is in the book. Blast, no not the first one, the last one, yeah, the last one, that's what he said.

What would happen if all comedy sketches were about project management? Wouldn't that be great? Not sure? Well, no need to worry. You no longer have to imagine it, here is a collection of twelve, yes count them, twelve comedy sketches and songs all about project management.

Through these twelve chapters, you will continue to follow the formative days in project management of our intrepid hero George Onaswell. You will have a fly on the wall view of his is trials, his tribulations and the characters that shaped his project management career.

"Continue?" I hear you say, did you not read When I Were a Project Manager? Get to it now! It's cheap and only a short read. Oh wait, please buy this one first,

I have a wife, children, and a really bad
fancy coffee habit to feed/quench.

Pick up your copy at amzn.to/2XBuj2U.

Coming Soon

Project Management: The Interviews

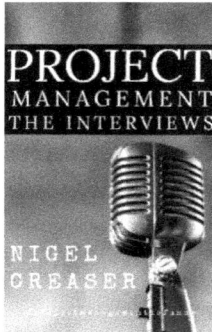

In addition to being a prolific author (Ed. Prolific? What? More like a lazy b*st*rd author) ...erhm... thanks Ed. Anyway, as I was saying as well as being an author, I host a small but perfectly formed Podcast called The Sunday Lunch Project Manager Podcast (anchor.fm/sundaylunchpm). I interview the great and the good of the project management world, from Lazy to Digital to the Doctor.

During one of the interviews, I thought a book transcribing the conversations might be of interest to people who prefer the written word rather than the spoken word. Though having started I discovered that transcribing was more difficult than I envisioned.

Sign up to the mailing list to get early notification of the release date or subscribe to the podcast where I will announce the availability.

www.nigelcreaser.com/mailinglist

The Sunday Lunch Project

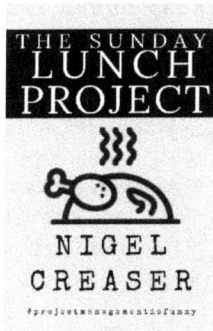

George Onaswell is a typical Project Manager and has been delivering projects for years, an adept stakeholder manager and Gantt chart guru, then he gets the most terrifying high-profile project of his career!

Cooking Sunday lunch for his prospective mother-in-law, head of the local Women's Institute and national roast dinner of the year champion four years on the trot.

How can he make it a roaring success while still delivering the thorny project he just got landed in the office?

The project charter is drafted for...

The Sunday Lunch Project!!!

What could possibly go wrong?

In all good bookstores (and rubbish ones) sometime after 2022.

Sign up to the mailing list to get early notification of the release date or subscribe to the podcast where I will announce the availability.

www.nigelcreaser.com/mailinglist

About the Author

Nigel Creaser, PMP, PMQ, PSM, PSPO is an experienced project and programme manager with over 20 years of varied project management roles, delivering multi-million-pound projects across a wide range of industries including national and regional government, financial services and telecoms and a former Director of Marketing for the Project Management Institute's UK Chapter.

He lives in North Shropshire with his wife and two daughters. When he is not mangling projects or being a husband and dad, you can find him on a judo mat trying to stay standing, trying to run a bit faster and a little bit further, getting a not unpleasant sound out of a guitar or tinkering with his motorcycle.

To get the latest updates why not pop along to www.nigelcreaser.com or sign up for the newsletter here to get them in your inbox.

www.nigelcreaser.com/mailinglist

www.ingramcontent.com/pod-product-compliance
Lightning Source LLC
Chambersburg PA
CBHW071643210326
41597CB00017B/2094